How Birds Fly

How Birds Fly

David Goodnow

Photographs by the Author

Illustrations by Robert J. Savannah

PERIWINKLE BOOKS INCORPORATED • COLUMBIA, MARYLAND

Copyright © 1992 by David Goodnow

All rights reserved. Except for use in a review, no part of this book may be reproduced by any means or used in any manner whatsoever without written permission from the publisher.

Published by Periwinkle Books Incorporated
Box 980, Columbia, Maryland 21044-0980

ISBN 0-9634244-0-8

Manufactured in the United States of America
Printed by Herlin Press Inc., West Haven, Connecticut

First Edition

To Peggy

When cities prod me with demands
Of many minds and many hands,
When life becomes a cry of bargains
In unassimilated jargons,
And men bewilder men with words
Suddenly I remember birds...

—Louis Untermeyer
Return to Birds

CONTENTS

CHAPTER 1

THE SEQUENCE IS THE SECRET

A bird takes to the sky.

We see a magnificent blur—then the repetitive wingbeat of the bird in full flight. But what took place during that tiny fraction of a second?

How did those flapping wings propel the bird aloft?

How does "thin air" suddenly support its weight?

How can the bird climb so swiftly?

How can feathers, bone and muscle conspire to master the sky with such grace and beauty—often with seemingly modest effort by the bird?

These questions are surely as old as humankind. They went without answers for nearly as long as we have shared the Earth with birds, because the unaided eye cannot unravel the riddle of bird flight. Too many things happen too quickly for us to decipher the process simply by watching birds fly. Bird flight remained a beautiful mystery until the invention of photography made it possible to record the *series* of individual wing motions that launch a bird into the air and keep it aloft.

Sequential photography makes it clear that bird flight is a remarkably straightforward process—a progression of steps in which the bird's wings move in a complex pattern that simultaneously generates upward *lift* and forward *thrust.*

A sequence of photos that captures a succession of wing positions *shows* what those wings are doing during each phase of the wingbeat as they support the bird and propel it through the air. I have found that a dozen sharp photographs can be worth twelve thousand words of anatomical terms and aerodynamic jargon to those of us who love birds

A powerful downstroke of its wings supports this herring gull and propels it through the air. The details of how bird wings accomplish the clever feat of generating lift and thrust are presented in Chapter 2.

The downstroke is complete; the gull begins to raise its wings in preparation for the next downstroke. The upstroke movement can be tricky to follow, because the bird hoists its inner wings first.

but are not engineers, physicists or ornithologists.

There are two ways to create sequences of wingbeat photos. The first is to photograph a bird in flight with a high-speed, motor-driven camera that records a series of successive wing positions (typically eight to twelve images per wingbeat) in a tiny fraction of a second. I used a sequence camera to film the herring gull on these pages and the Canada goose that stars in Chapter 2.

The second method is to "assemble" a wingbeat sequence by repeatedly photographing a flying bird in a specially equipped studio. I used a conventional 35-mm camera and a bank of electronic flash units to make the photos of ducks and owls that follow.

Both techniques demand photographic gear that can "freeze" the flapping of the bird's wings. My high-speed sequence camera has a shutter that remains open for only one three-thousandth of a second. The electronic flash units I use with my conventional cameras produce exceptionally brief pulses of light (typically lasting less than one nine-thousandth of a second).

Birds are among my favorite subjects; over the years, I have taken

The gull keeps its "hands" pointed downward (and the feathers on its wing tips spread apart) to minimize the air resistance as its wings move upward and backward during the recovery stroke.

With a quick flick of its wrists, the gull raises its hands at the end of the recovery stroke. When it fully straightens its wings, the bird will be ready to begin the downstroke of its next wingbeat.

thousands of photographs of birds aloft. I chose the flight sequences and individual photos that appear in this book because they help answer—in the simplest possible way—the question: how do birds fly?

Bird flight is a vast topic; I have not tried to cover its many different aspects and technicalities in this book. My purpose is to help you grasp the essentials of flapping flight—the common-denominator principles that apply to the way virtually all birds fly, be they big or small, fast or slow, familiar or exotic. Master these fundamentals, and you will find it easy to understand the twists and variations presented in lengthier reference works on avian flight.

You will also experience a heightened sense of wonder when you see birds fly. The actual mechanics of bird flight are much more interesting than the familiar (but incorrect) notions that birds somehow "swim" through the sky or "beat down" the air to stay aloft.

Bird flight is certainly a miracle. But thanks to the alert eye of high-speed sequential photography, it is no longer a mystery.

CHAPTER 2

WINGS SWEEPING THROUGH THE AIR

A bird flies for the same reason that an airplane flies: they both produce *lift* and *thrust*.

The bird rises when sufficient *lift* is generated by its wings to overcome the downward pull of *gravity*.

The bird moves forward in response to *thrust* (similarly generated by its wings). A flying bird also experiences the rearward tug of *drag*—a combination of forces that work to slow the bird as its body pushes through the air and its wings produce lift.

Of course, an aircraft designer enjoys the luxury of using two different kinds of machinery to generate upward and forward forces:

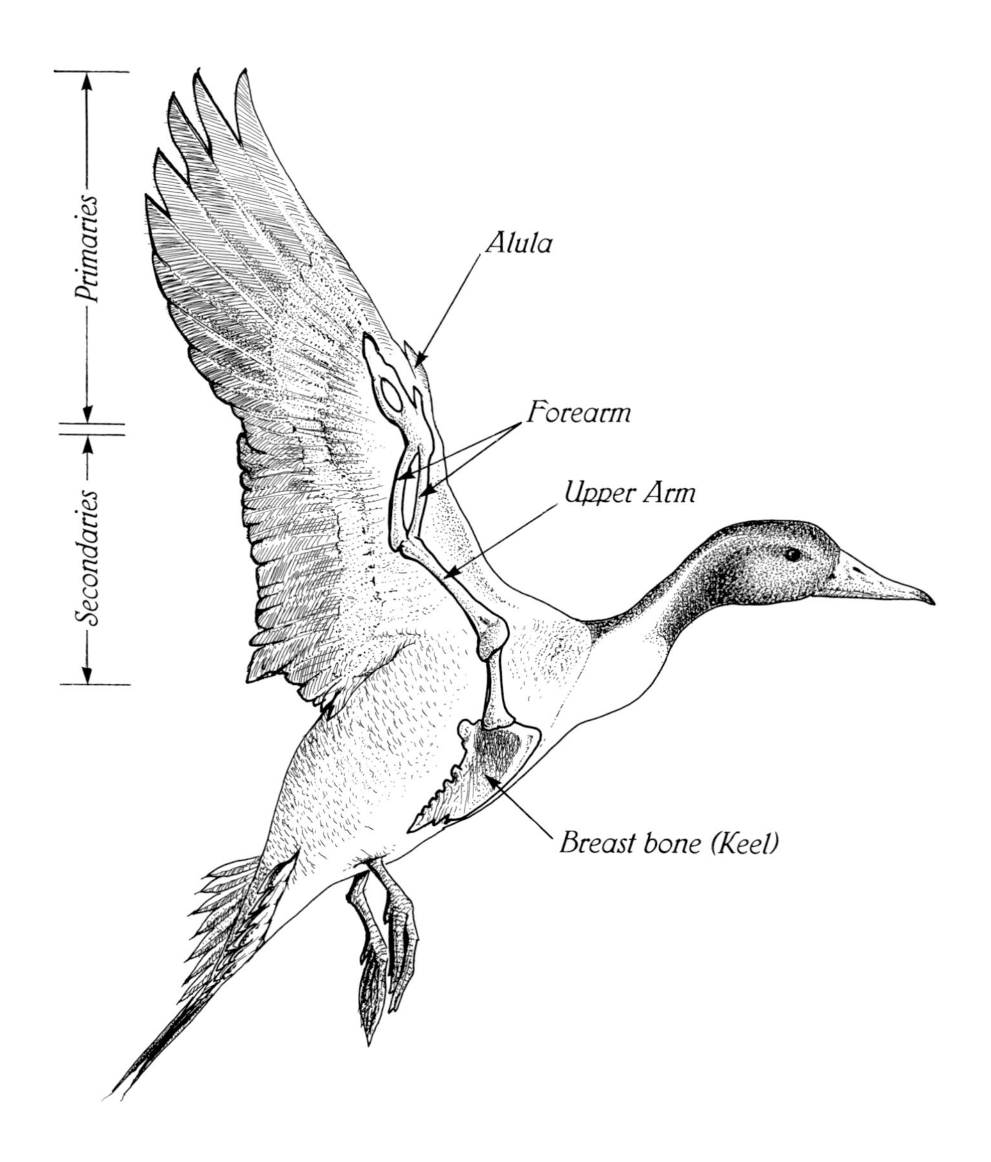

wings to create lift and engines to produce thrust. A bird makes do with one mechanism: a pair of wings made of skin, bones (that resemble the bones inside a human arm and carry similar names) and a variety of feathers. The bird must simultaneously generate lift and propulsive thrust by using muscle power to flap its arms and hands.

A pintail drake attempting a dramatic near-vertical take-off has just completed an energetic downstroke that simultaneously produced lift and thrust—enough to propel the bird upward at steep angle.

But don't be deceived by what you think you see. Birds do not "beat down" the air with their feathers when they fly. Flapping flight requires that a bird's wings sweep *through* the air in a highly complex pattern. Motion through the air is essential. No wing—be it bird or airplane—can create lift unless air is flowing past its elegantly curved surfaces.

We will prove the point by watching a Canada goose fly.

1. The Downstroke Begins

The Canada goose makes an excellent "model" to demonstrate flapping flight because it is a relatively large bird that swings its long wings comparatively slowly. Each stage of its impressive wingbeat is pronounced and easy to interpret.

This particular Canada goose is poised to begin a downstroke. Try to envision the bone structure inside the wing: the upper arm, the forearm, the wrist, the hand (with its various bones fused together to support the bird's primary feathers). Observe that the raised wing has three different kinds of *flight feathers*:

- The ten large *primaries* attached to the hand at the end of the wing
- The *secondaries* behind the forearm on the trailing edge of the wing
- The patch of *tertiaries* that close the gap between the secondaries and the bird's body

Convenient points of reference are essential when you study wing movement. We will use the tip of the bird's bill and two highly visible parts of the wing: the wing tip and the wrist (visible as a bump on the front of the wing). In this photo, the goose's wing tip is about as far away from the bill as it will get.

2. The Downstroke Continues

At first glance, it is difficult to see any difference between this picture and the previous photo. Nothing much seems to have happened.

In fact, our Canada goose's "propellers" have swung into action.

Look closely at those large primary feathers (also called *pinions*) on the tips of the wings. Notice how they have begun to twist in response to the air rushing past their wide rear vanes.

Each of those primary feathers has twisted just enough to bite into the air—and generate forward thrust—as the wing moves downward. In other words:

- A bird's primary feathers function much like the propeller blades on a prop-driven airplane.
- The hand sections of the wings that hold the primary feathers work much like propellers—they sweep the propeller blades (the primaries) through the air to generate the thrust that propels the bird forward.

What holds the goose aloft while its hands are making thrust? During flapping flight, most of the lift to support the bird is provided by the inner wing (the section from shoulder to wrist that carries the secondary and tertiary flight feathers). The inner wing maintains its lift-generating capabilities throughout much of the wingbeat cycle.

3. The Downstroke Continues

Flying is the single-most strenuous activity that animals perform. A lot of goosepower is necessary to sweep those broad, fan-like wings through the air as the Canada goose strives to gain altitude.

Bird wings and airplane engines must both work harder during takeoff and climb than during level flight. Once a bird reaches cruising altitude, it can throttle back the powerful wingbeats that launched it into the air and begin the seemingly effortless "sculling" of level fight.

A bird's "engines" are the hefty pectoral muscles anchored to its *keel* (breastbone). A big pectoral muscle (*pectoralis*) forces the wing downward; a weaker pectoral muscle (*supracoracoideus*) powers the upstroke. Unlike most other vertebrates, birds do not have strong back muscles to articulate their limbs—for a good reason. Pectoral muscles can account for as much as a third of a bird's weight. Locating heavy muscles on its breast lowers the bird's center of gravity and improves flight stability.

Getting back to our goose—observe that its wing tips and wrist have both moved closer to its bill, because wings are being pulled forward by the thrust of the "propellers." The primary feathers (angled sideways to bite the air deeply on the downstroke) have begun to bend noticeably.

4. The Downstroke Continues

You can almost feel the flight stresses developing inside the wings of our Canada goose as the bird continues its powerful downstroke. Its wrist is nearly fully extended and the wing bones are lined up to form a stiff "spar" (like the supporting structure inside an airplane wing).

At this point in the wingbeat cycle, it is possible to see the wing's curved cross-section—the familiar arched shape that gives an *airfoil* the ability to generate lift. You also have a bird's eye view of the primary feathers—the bird's "propeller blades"—at work.

Primary feathers are, of course, non-symmetrical: most of the feathery *vane* lies on one side of the stiff *shaft*. It is the shaft edge of each primary feather that bites into the air on the downstroke. The outermost (first) primary is connected to—and controlled by—a bird's single finger tip. The bird can articulate this feather and use it to guide the air flowing past the wing tip. This becomes important during slow flight.

The other feathers on the wings are called *coverts*. Small covert feathers on the leading edge and larger coverts on the top give the wing its airfoil shape. Observe how the coverts on the top of the wing lie flat like the overlapping shingles on a roof—a measure of the smoothness of the airflow across the wing's main lift-producing surfaces.

5. The Downstroke Continues

The Canada goose's wings are driving downward at a high rate of speed (there was a significant change in wing position during the small fraction of second between this and the previous photo). The wing tip has moved level with the base of the bird's neck; the wrist is approximately in line with the bird's breast; the primary feathers (still biting strongly into the air) are bent forward under the strain of producing the thrust that is propelling the goose ahead.

Bird feathers are made of a tough, resilient material called keratin; primary feathers are particularly strong and stiff. Even so, the primaries of larger birds flex to a remarkable degree during flapping flight.

But note that the goose's wing has maintained its overall shape.

Humans have to use muscles to keep their arms outstretched and stiff, but the fewer joints inside a bird's wing are arranged so that they are naturally inflexible in the vertical plane, the direction of maximum stress during the downstroke. The wrist and elbow joints allow the wing to fold sideways against the bird's body when it is not flying, but they do not permit much up-and-down flexing of the wing sections when the arm is fully extended. Consequently, a bird need not waste muscle power to hold its arms straight.

6. The Downstroke Continues

The tips of our Canada goose's wings now reach nearly to its bill—a sizable stretch, considering the length of a goose's neck. It is worth recalling one more time that the wings are being *pulled* forward by the thrust generated by the primary feathers as they bite into the air. (The wing tips of many birds are actually drawn past their heads when their "propellers" are generating maximum thrust—during takeoff and climb, for example, or when the bird decides to accelerate.)

Of course, the goose's wings traveled a much greater distance *through* the air, because the bird flew several feet forward while its wing tips were slicing downward. That's why the wings were able to generate lift and keep the bird aloft.

Now here is a fact that may surprise you: the inner sections of both wings stayed relatively level—and positioned to produce lift—during much of the downstroke.

It's true!

The swipe of the goose's hands—the half-revolutions of its "propellers" to produce thrust—are responsible for most of the wing motion that you have seen so far. The goose's inner arms definitely moved downward, but only through a relatively small angle that maintained the wings' ability to generate lift.

7. The Recovery Stroke Begins

Our Canada goose has finished its strenuous downstroke. Observe the primaries at the end of the wing; they are producing much less thrust and have begun to unbend. The goose's momentum will keep it coasting through the air—its inner wings generating lift—even though its "propellers" have stopped making thrust.

The recovery stroke—the second half of a wingbeat—is more difficult to understand than the downstroke because of the tricky goings-on inside the bird's wing. If you compare this picture with the previous photo, you will see two things happening to start the upstroke:

- The inner wing has begun to move upward
- The hand has begun to flex at the wrist

And therein lies a tale.

Hummingbirds (and many insects) have wings that generate thrust and lift during both the downstroke and upstroke. That's why they can hover. But for most other birds, the chief objective is to get the wings back into a raised position—ready to begin another downstroke—as quickly as possible, without causing a significant loss of lift.

The avian solution involves a slick mechanical trick. The bird raises its inner-wing sections first, then it flips its hands upward—with quick flicks of its wrists—to complete the recovery stroke.

8. The Recovery Stroke Continues

Now, the downward rotation of our Canada goose's hand is even more apparent (observe the pronounced flex of the wrist). The bird's inner wing sections, which have risen considerably since the previous photo, are also moving backward—additional evidence that the bird's "propellers" are no longer producing the thrust that pulled the wings forward during the downstroke.

The small pectoral muscle that drives the upstroke is perhaps only one-tenth as heavy as the large pectoral muscle that powers the down-stroke, yet the recovery stroke takes less time to complete than the downstroke. The are three chief reasons:

- The front edge of the inner wing adopts a greater upward tilt during the recovery stroke, so that the forward motion of the bird through the air helps to push the whole wing up and back.
- A partially folded wing offers significantly less air resistance than a rigid wing.
- The bird's primary feathers align themselves to further reduce air resistance during the upstroke.

If your eye is especially sharp, you may notice that primaries seem a bit farther apart in this photo. They have already begun to twist to an "open" position that can pass through the air more easily as the wing moves up and back.

9. The Recovery Stroke Continues

At this stage of the wingbeat, our Canada goose seems to be flying with "squared" wings—the tips of its primaries aimed generally downward; its inner wings parallel to the ground.

Keep in mind that a bird does not bend its hand downward during the recovery stroke. It can't. The wrist joint only permits rotation in the same plane of the hand. (To visualize this kind of motion, think of brushing crumbs off the top of a table.)

The goose's ten primary feathers point downhill, the result of a complex mix of anatomy and motion:

- Other muscles that work only during the recovery stroke have angled the leading edge of the wing upward.
- The goose has flexed its wing at both elbow and wrist, and has lifted its wing at the shoulder.
- The bird has rotated its hand downward (and slid the ten primaries closer together like the sticks and leaves of a hand-held fan).

This photo, more than any other in the sequence, shows how big the hand is in relation to the whole wing. If the large "propeller" surprises you, consider this: Although *lift* is the essential ingredient of flight, *thrust* is the key to generating lift during flapping flight. Thrust is what pulls the wings through the air.

10. The Recovery Stroke Continues

Our Canada goose is making things happen briskly at this stage of its wingbeat:

- The bird's wings have risen significantly; its wing tips have moved backward a notch.
- The bird's inner wings now resemble air scoops; they are angled upward to catch the air rushing past—and gain a free push to their eventual raised position.
- The bird's elbows have flexed upward, lifting the wrists (once again, its wrist is on a level with its neck).
- The bird's hands have rotated downward several more degrees and fanned the primary feathers closer together.
- The primaries have twisted almost a quarter-turn to "open" like Venetian blinds, reducing the air resistance that would otherwise slow the upstroke (the primaries on the goose's right wing provide the best view).

Some birds are able to generate thrust during the recovery stroke. They position their hands so that their primaries bite "backward" into the air on the upstroke. This maneuver is used primarily during slow flight, when thrust (aimed upward instead of forward) can compensate for diminished lift and help to support a bird in the air.

11. The Recovery Stroke Continues

Our Canada goose has lifted its wrists above its body—they are almost as high as they were 11 photos ago, at the start of the downstroke. The bird is continuing to hoist its wings upward, its two "propellers" completely out of service:

- The wrist joints are flexed downward
- The hands are relaxed
- The primary feathers are aligned to minimize air resistance during the upstroke

An X-ray of the goose's wing would show the bones arranged in a rough "Z," with the elbow and wrist joints bent, and the inner arm folded between the outer arm and the hand.

This wing position may strike you as a bit perilous; a time of inefficient airflow and enormous drag. You may wonder why the goose's body does not sag downward.

Actually, the bird's up-angled inner wings continue to generate lift throughout the upstroke (and also some drag, which pulls the wings backward).

Lack of thrust is not a significant problem, either. The recovery stroke is over so quickly that the bird does not lose much forward momentum.

12. The Recovery Stroke is Complete

The Canada goose's wing seems to have undergone a remarkable transformation since the previous photograph.

Look again.

The chief difference between the two wing positions is that the goose flicked its wrists during the brief interval between the successive frames:

- The wrists are rotating upward as the wings extend, to open the "fans" of primary feathers and prepare the hands for the next downstroke.
- The primaries are spreading apart and twisting closed, readying themselves to function as "propeller blades."
- The extending wings probably produce some forward thrust when the backs of the hands and inner wings push against the air, much like a swimmer shoving-off from the side of a swimming pool.

Observe that the tips of the primary feathers are blurred—a good indication of how quickly the goose is rotating its wrists. A shutter speed of one three-thousandth of a second was not quite fast enough to "freeze" the rapid motion.

CHAPTER 3

BERNOULLI AND THE BIRDS

A vast abyss of time separates *Archaeopteryx*—the primitive bird that had much in common with small dinosaurs—from the birds of today. The finished product of all these eons of evolution is a large family of creatures of consummate beauty, fantastic coordination and superbly efficient utilization of energy.

Birds come in many different shapes and sizes—each best suited for a specific function in nature. But all birds must apply the same principles of aerodynamics when they fly. It is pretty much a case of similar strokes for different folks.

When a properly shaped airfoil—be it a bird wing or an airplane wing—moves through the air, it generates the upward force we call lift. The basic idea is quite simple.

A wing slices the air into two air flows, one passing atop the wing, the other passing beneath. The wing's curved shape forces the upper mass of air to flow faster, because it must travel a longer distance than the air sliding below the wing.

But as Daniel Bernoulli (a Swiss physicist who lived during the 18th Century) observed, when flowing air accelerates, its pressure *decreases.* Consequently, the pressure of the air flowing above the moving wing is *lower* than the pressure of the air flowing beneath the wing. These pressure differences push the wing upward. The "Bernoulli Effect" is the source of the force we call lift.

Controlled lift is the crux of flight, whether by an airplane or a bird. A plane or a bird climbs when the lift generated by its wings exceeds its

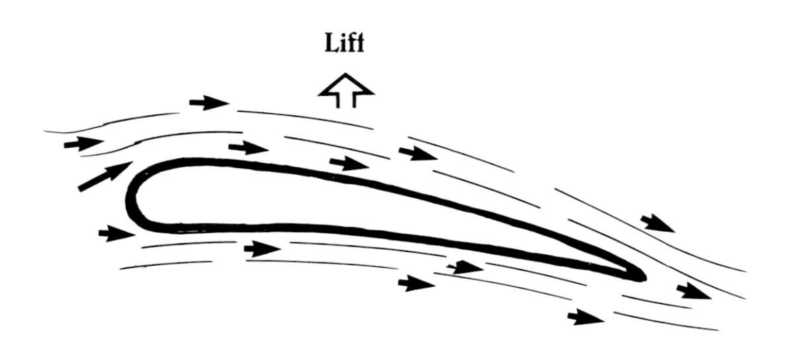

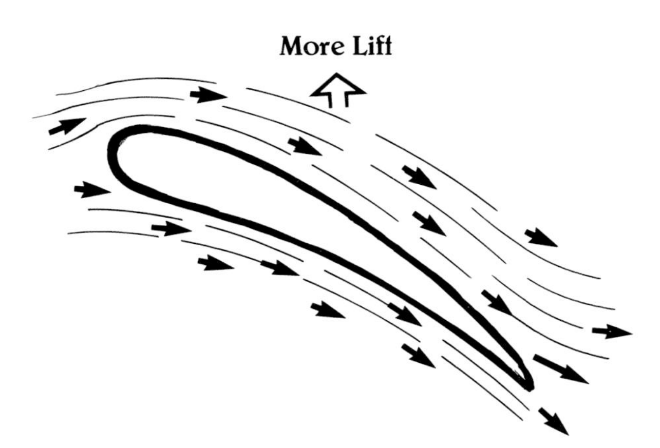

weight. A plane or a bird glides to a gentle landing when weight *slightly* exceeds lift (a major discrepancy between lift and weight will cause a crash). A plane or a bird flies at a steady altitude when lift exactly balances weight. A plane or a bird will "bank" when the greater lift produced by one of its wings induces a lean toward the other wing.

The amount of lift created by a wing depends on many factors, but three of the most important are:

1. The wing's "angle of attack"—the greater the angle at which the wing cuts through the air, the more lift…up to a point. If an airfoil meets the air at too steep an angle, the air flowing over the top breaks loose from the surface and burbles. The wing *stalls*; it loses its lift and stops flying.
2. The wing's shape—a thicker curvature generally increases the lift.
3. The speed of the wing through the air—all things being equal, the lift generated by an airfoil is proportional to the square of its speed through the air. Doubling the airspeed boosts lift by a factor of four; halving the airspeed decreases the wing's lift to one-quarter of its original value.

A great horned owl, flying slowly, is capable of extremely precise control of lift, thrust and drag. The bird's wing is an exceptionally complex airfoil that has different angles of attack at various points along its length.

Birds and pilots manipulate all of these factors to control their flight and prevent stalling. But birds do a better job. A bird has an awesome ability to vary the shape of its wings with a twitch of a muscle or a flick of a feather. A bird can accomplish changes in attitude far more subtly than any human, who must settle for ailerons, flaps and other mechanical devices.

Pilots daydream about being "one with their aircraft." Birds are.

This great horned owl has just begun its downstroke. The leading edge of its inner wing is aimed upward for a greater angle of attack; its hand is angled downward to give the primary feathers a deep bite into the air. The tips of the primaries are "fluffy." This unusual modification transforms an owl into a "stealth" flyer by muffling the sound of the downstroke.

Wings high, a great horned owl is ready to launch itself from its perch. We have an excellent look at the different feathers on the bird's wing: the primaries at the end of its hand, the secondaries along the trailing edge of the inner wing, the coverts along the leading edge and the tertiaries that fill the gap between the wing and the body.

Our great horned owl rotates its wings at the start of a downstroke. Its primary feathers are twisted to bite into the air and generate thrust.

The secondary feathers on this great horned owl's left wing flutter in the breeze during an upstroke. The bird is about to extend its hands.

Its alulas raised, its wing and tail angled to generate maximum lift, the great horned owl completes a long-reaching downstroke.

An osprey applies its "air brakes" to make a pinpoint landing on a small perch. Its primary feathers face the airflow to increase drag.

An osprey flies to a perfect low-speed landing. Its alulas are raised to prevent stalling, its primaries and tail feathers act as "air brakes."

A moment after taking off into the wind, this osprey is about to complete a recovery stroke by flicking its wrists to open its hands.

A ruffed grouse shows its impressive tail feathers as it makes a characteristically explosive takeoff. The recovery stroke is nearly complete.

The ruffed grouse has relatively small rounded wings, which are appropriate for its woodland habitat. Here, the bird begins a vigorous downstroke.

A blue wing teal, climbing at a steep angle, completes its recovery stroke by snapping its "propellers" into position for the next downstroke.

The blue wing teal, now near the end of a downstroke, has its alulas raised to stabilize the air flowing across the tops of its inner wings.

A black duck, climbing nearly vertically, reduces drag during its recovery stroke by drawing its folded wings close against its body.

The black duck has inverted its hands during the upstroke of its wingbeat—a flight technique that provides additional lift and thrust.

A wood duck ends an exuberant recovery stroke by extending its wings behind its body. Its primaries are still "open" to reduce air resistance.

Poised to begin a new downstroke, the wood duck shows off its heavily loaded wings. They are optimized for high speed flight and fast climbs.

The wood duck's small wings require powerful downstrokes to keep the bird aloft. Note how the primaries bend under the strain of making thrust.

CHAPTER 4

AVIAN AERONAUTICAL ENGINEERING

Birds are extraordinary "flying machines."

- The American golden plover migrates nonstop from Labrador and Newfoundland to the northern coast of South America, a flight of more than 2,500 miles.
- An albatross was once recaptured after having flown 3400 miles over open ocean in eight days.
- The Arctic tern may fly a total 22,000 miles on its annual migratory round trip from the Arctic to the Antarctic, and back.
- Many hummingbirds choose a direct route from the United States to their winter homes in Mexico—a 500-mile nonstop flight across the Gulf of Mexico.
- Homing pigeons can fly for hundreds of miles at airspeeds in excess of 50 miles per hour.

Bird anatomy shows a high degree of specialization—always for the purpose of flight. Birds are heavy in areas that are essential to flight (pectoral muscles account for a third of the weight of some birds) and light in areas that are not (birds have very thin skulls). They are equipped with unusual breathing organs that do a superb job of extracting oxygen from the air. And, of course, birds have multitudes of specialized muscles to control their feathers. You can see many of them at work on the ground when a peacock rattles its plumage.

The relative sizes of the large pectoral (the muscle that powers the downstroke) and the small pectoral (the muscle that drives the upstroke) depend on how a bird is designed to fly and take off. Fast flapping birds,

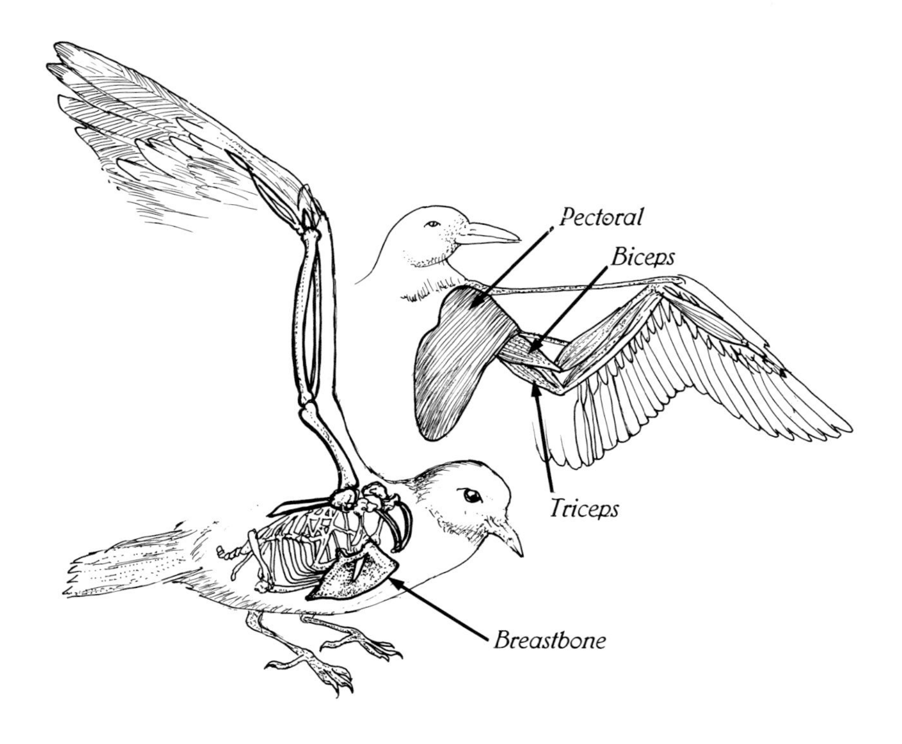

Birds have large breastbones to anchor their major flight muscles, the heavy pectoral muscles that drive their wings down and up. The pectorals fill the space between the breastbone's keel and the bottom of the wing.

which can produce enough thrust and lift on the upstroke to takeoff vertically and climb at steep angles, need comparatively heavy small pectoral muscles (perhaps a third as large as the big muscle). Slow-flapping birds have small pectorals that may be only a tenth as heavy as their large muscles.

Flight is a strenuous activity; a bird's "power plant" is designed to deliver a steady flow of energy to the bird's muscles. The sources of this energy are the chemical reactions that "burn" the food the bird eats in the oxygen it breathes.

Birds have a rather unusual breathing apparatus. Inhaled air flows through a system of *airsacs*, a capacious network of bubble-like, membranous bladders located throughout the bird's body. There are several

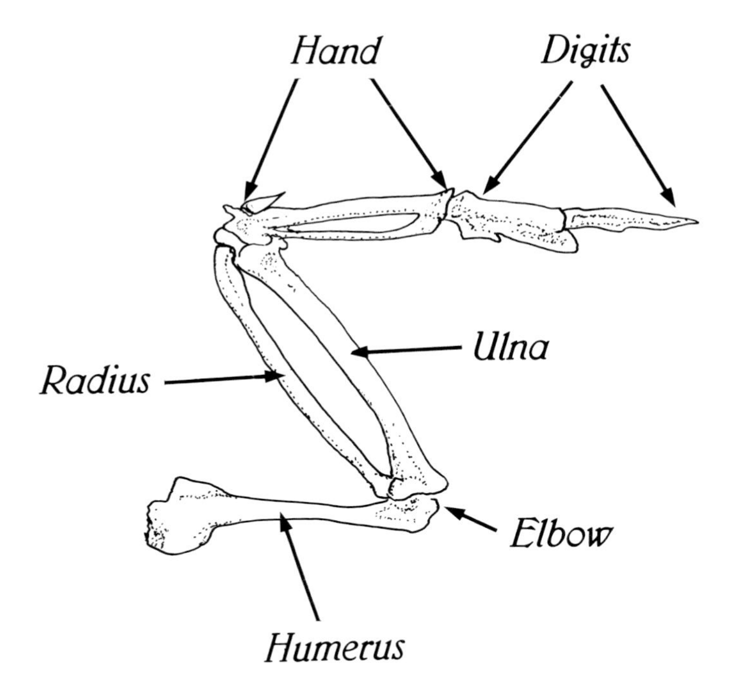

The bones inside a wing have familiar names, but many are significantly different than their human counterparts. Note the large overall length of a bird's "propeller"—its whole hand, including the "digits."

theories that try to explain how airsacs improve the ability of bird lungs to extract oxygen from the air. One theory holds that the way a bird breathes encourages a steady, one-way flow of air through its lungs.

At first glance, the innards of a bird's wing resembles a human arm. In fact, a bird's *humerus* (upper arm) is considerably shorter than its human counterpart, its hand is comparatively longer and its wrist is simpler, with fewer bones.

The wrist and elbow joints allow the wing to fold sideways against the bird's body, but they do not flex in the vertical plane when the wing is fully extended. Consequently, the bone structure provides a high-strength "wing spar" that will hold the wing rigid during a powerful downstroke.

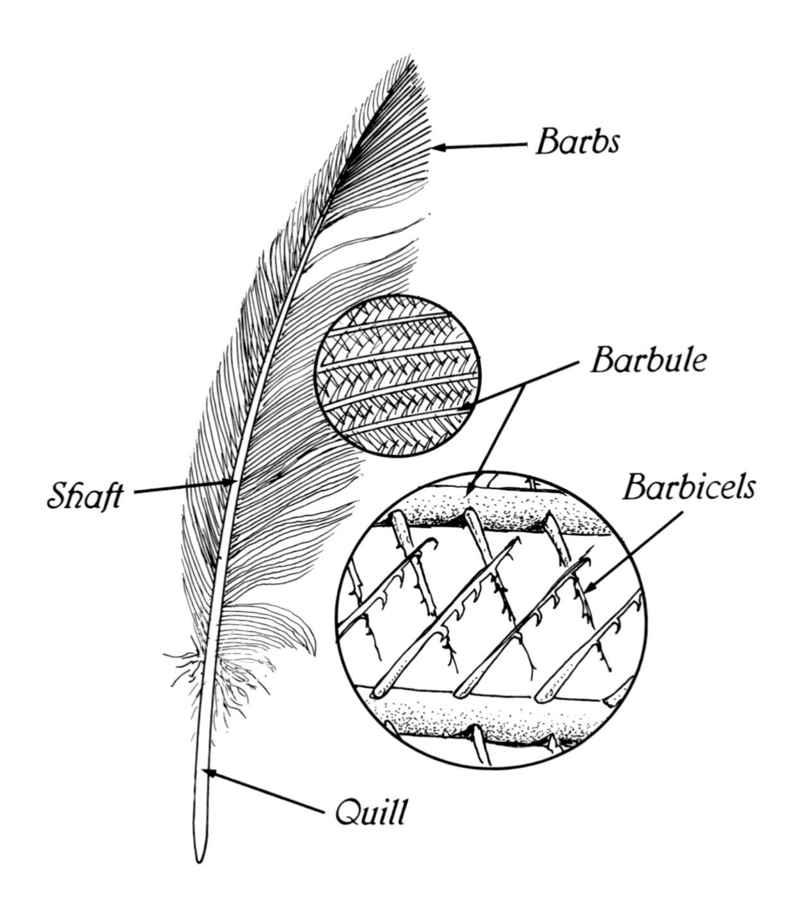

A feather is a surprisingly complex device—an assembly of more than a million individual parts. Tiny hooks on the interlocking barbules tie the barbs together to create tough, resilient, easily maintained vanes.

Our Friends' Feathers

Feathers are among nature's most ingenious creations—and absolutely essential for bird flight. A bird's wing is so light for its size because it consists mostly of feathers and the skin that supports them.

A feather is a dead structure—an out-pushing of the epidermis, much like bone, horn, or hair—made of a tough, resilient material called keratin. Feathers are strong, light and able to withstand the rigors of flight and the

When the barbs of a feather are locked together side-by-side by the barbules, they create a durable "fabric" web that is essentially airtight. Most birds replace their feathers in a periodic (usually annual) molt.

vagaries of climate. A feather consists of a *shaft* that supports the vane or web. The base of the shaft—the *quill*—is imbedded in the bird's skin.

- *Barbs* extend on both sides of the shaft diagonally upward toward the tip of the feather like branches growing on a tree trunk.
- Hooked fibers, called *barbules*, interlock each barb to its neighbor.
- Even more minuscule *barbicels* hook on to adjancent feathers to further stabilize the wing surfaces.

The barbs that comprise the vane can be easily rejoined when pulled

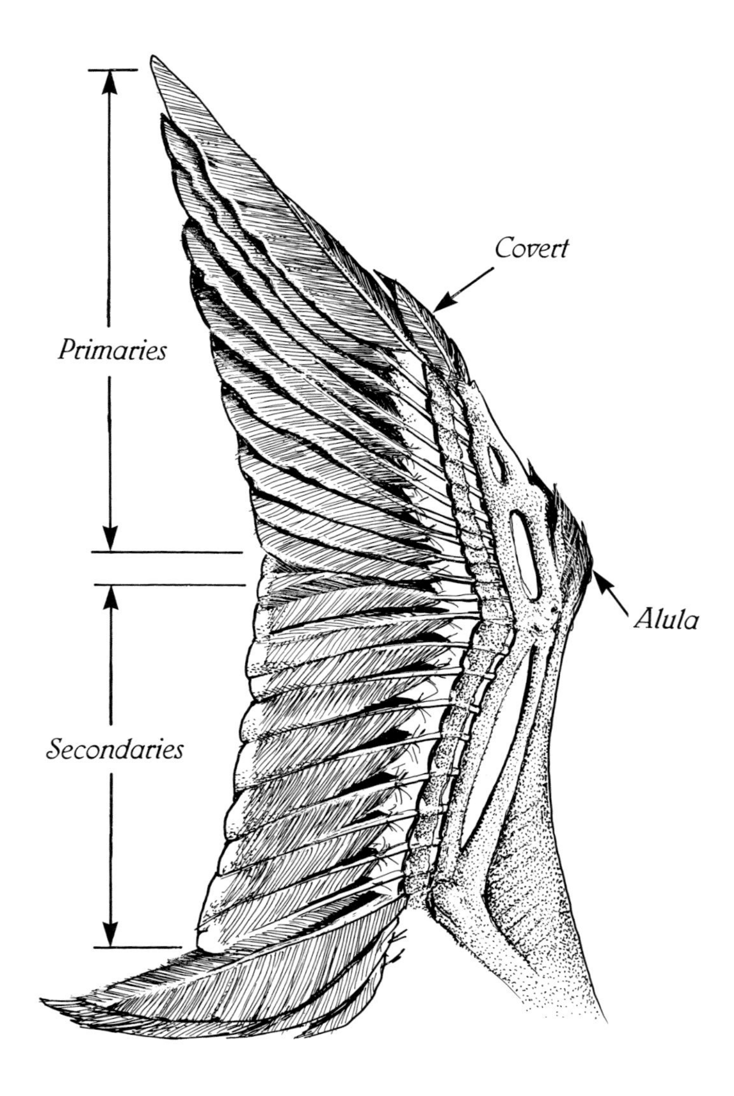

apart, rather like Velcro. Disheveled plumage is quickly set right when the bird preens itself. Birds are equipped with several kinds of feathers:

- *Primary* feathers—from the wrist to the wing tip—produce the thrust that pulls the bird through the air. Most birds have ten primary feathers. They are sharp-pointed and asymmetric—the vane on the

The lift generated by this raven's vigorous downstroke has placed its handful of primary feathers—its "propeller blades"—under high stress. Primaries bend to a surprising degree during takeoff and climb.

feather's leading edge is narrower than the vane on the trailing edge.

- *Secondary* feathers cover the trailing edge of the wing, from the wrist to the elbow. They constitute a major portion of the wing's lifting area.
- *Tertiary* feathers fill the space between the elbow and the bird's body.
- C*overt* feathers cover the wing to provide the curved airfoil shape. The coverts on the leading edge are smaller than the coverts on top.
- *Contour* feathers grow on the bird's body to create the streamlined body shape necessary for efficient flight.
- *Tail* feathers are similar to primaries and secondaries, but are usually symmetrical.
- *Down* feathers are fluffy because they have are no barbules to lock the barbs together. They serve to insulate the bird's body.

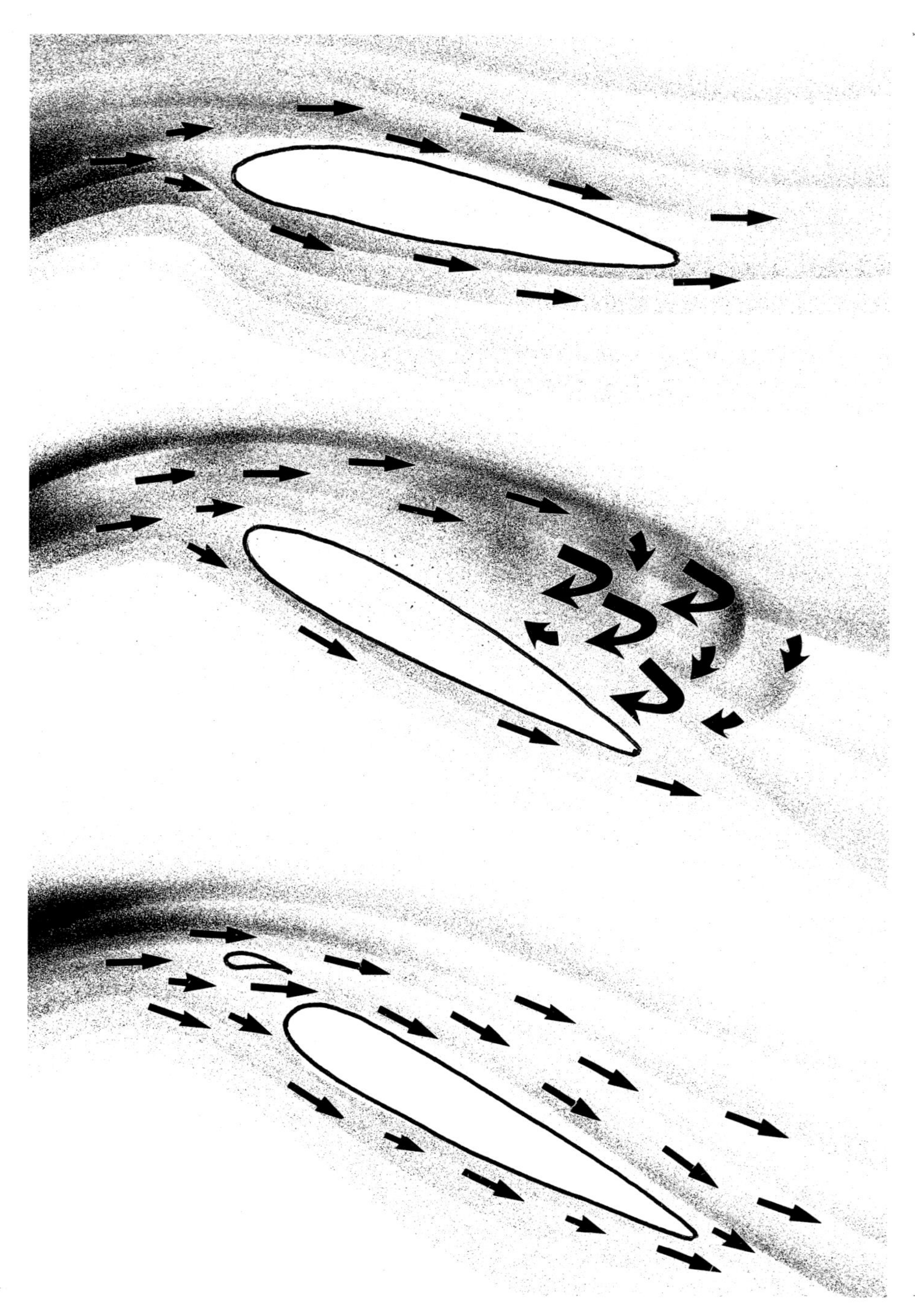

Air flows smoothly above and below an airfoil flying at a moderate angle of attack (top). But the air above the wing will break loose (middle) if the angle of attack is made too steep—the wing stalls, and stops flying. A slot near the wing's leading edge (bottom) can help guide the airflow over the wing and prevent stalling during low-speed flight.

CHAPTER 5

LOW AND SLOW

Slow flight is a challenge for both birds and aircraft. Slow speed through the air means that wings produce less lift. On the other hand, slow flight cannot be avoided during takeoffs, landings and complex maneuvers near the ground or in confined spaces.

Birds and pilots can compensate for reduced lift by increasing the angle of attack of their wings. But too high an angle of attack will cause the smooth airflow over the top of the wing to break loose and burble. The wing *stalls*; it stops generating lift.

Man-made wings have various devices for generating additional lift at slow speeds and preventing the onset of stall. Virtually all airplanes are equipped with *flaps* on the trailing edges of their wings. Flaps increase the effective curvature of the wing. Some designs also increase the area of the wing by sliding downward and outward from the trailing edge.

Aircraft drop their wing flaps during landings and takeoffs; they can increase the available lift by more than 60 percent. But because flaps also significantly increase drag, they are usually raised when the aircraft is climbing or cruising.

Many sophisticated aircraft are equipped with *slots* on the leading edges of their wings. The slots help to guide the air moving across the upper surfaces of the wings, which reduces burbling and prevents stalls.

Bird wings make use of the very same aids to slow flight—usually all at once, in an extremely efficient manner. When landing or taking off, a slow-flying bird will have its wings upswept at a sharp angle of attack, its tail is outspread and lowered (in lieu of flaps) and the *alulas* on its wings

The alula, located on the leading edge of the wing, is controlled by the bird's thumb. It "bleeds" air over the top of the wing to reduce stalls during high-angle-of-attack flight. This wing belongs to a Canada goose.

raised. The *alula,* or false wing, is a "winglet" made of three feathers that is controlled by the bird's thumb. The bird uses each alula to guide and stabilize the air flowing across the top of the wing, which permits the wing to fly at a higher angle of attack without stalling. Use of the alulas is particularly noticeable when a bird is landing and needs maximum lift just before touchdown.

This combination of increased angle of attack, tail "flaps" (which can add an amount of lift equivalent to increasing a bird's wing area by 20

This tern's large tail provides additional lift during slow flight. It also helps to prevent stalling when the bird flies with its wings at a high angle of attack. The narrow gap between the front of the tail and the trailing edge of the wing serves as a slot that encourages air to flow more smoothly over the top of the inner wing.

Slow flight close to the ground requires additional thrust to compensate for the greater drag of high-angle-of-attack flight and "flaps down" maneuvering. This tern is beginning a vigorous downstroke, with its wings hoisted high above its head. The bird's primary feathers are bending noticeably under the strain.

This frame was made a few hundredths of a second later than the photo on the opposite page. The tern has commenced a left turn, "raised flaps" and begun to accelerate into normal flapping flight.

percent) and the alula anti-stall device allows a bird to land very slowly, with great precision.

When a bird raises its tail during low-speed flight, the gap between the front of the tail and the trailing edge of the wing may also serve as a slot that helps to smooth the flow of air over the inner wing. Many birds create slots near the leading edges of their wings by extending their first primary feathers. (The first primary is attached to the index finger, and is highly adjustable.) Large soaring birds have widely spaced primaries that act as

This vulture has wings designed for low-speed soaring flight. Its primary feathers are "emarginated" (reduced in width near the tips) to create wing-tip slots that reduce the tendency to stall at high angles of attack.

wing tip slots to control *tip vortex.* This is the burbling, whirling of air at a wing tip that can spread inward along the bird's wing, causing unstable airflow and consequent loss of lift.

Incidentally, tip vortex may be one reason that geese fly in the classic "V" formation that is one of nature's most stirring sights. The vortex spiraling backward from one bird's wing tips may provide a bit of free lift for the bird next in line.

Only the leader loses out. Watch a flock of Canada geese on migration,

An osprey making a slow approach to a landing uses several techniques to boost lift. The bird has increased the angle of attack of its wings, "dropped its flaps" and raised its alula to improve aiflow over its wings.

and you may see the leader drop back in the formation, to relinquish the strenuous point position to another bird.

Aircraft designers generally try to eliminate as much drag as possible with streamlined shapes and smooth surfaces. Birds used both of these techniques first. But there is one occasion when this usually unwanted force is essential: Drag provides the "brakes" that slow birds and aircraft so that they can land at reasonable speed.

Heavy birds must shed lots of momentum before they can touch

This osprey, coming in for a landing, is using its wings as "air brakes" to increase drag and reduce its airspeed. The ruffled feathers on the bird's back demonstrate that its wings have begun to stall.

down. The bigger the bird, the more pressing the problem. The challenge becomes even more acute when big birds want to land on small perches. Their solution is to transform their wings and tails into feathery air brakes that present large, high-drag surfaces to the on-rushing air.

Our osprey has maximized the effectiveness of its "air brakes" by extending its wings, holding its tail straight and rotating its primary feathers to face the airflow. The bird is steering by varying the relative drag produced by its wings; note how the primaries on the left wing are rotated downward to reduce drag, which will swing the bird's body to the right.

1

The gull unfolds and extends its wings. Air flutters past its large primary feathers, which are angled to reduce air resistance during the upstroke.

⇒

5

Looking much like an aircraft, the gull lifts off. Its primaries—its "propeller blades"—are bent under the load of the thrust they are producing.

⇐

Aloft in Half a Wingbeat

A flapping-flight takeoff begins with an exuberant downstroke that launches the bird skyward. The herring gull in this high-speed photo sequence makes the chore seem effortless. The bird was startled by the unfamiliar whirring sound of my remote-controlled sequence camera. It made a split-second decision to fly away.

Because the gull has relatively short legs, its initial downstroke cannot reach below its body (its wings will strike the ground). Consequently, the bird must maximize the duration of the downstroke—and the lift the downstroke generates—by starting with its wings high above its head.

The gull's takeoff toward the camera provides an excellent view of the bird's inner wing. Observe how the covert feathers on the leading edge and the secondary feathers on the trailing edge are arranged to create an elegantly curved airfoil.

Note also how the gull's lift-producing inner wings tilt upward, while its thrust-generating hands ("propellers") tilt downward. These alignments are built into the bone structure of the wing.

6

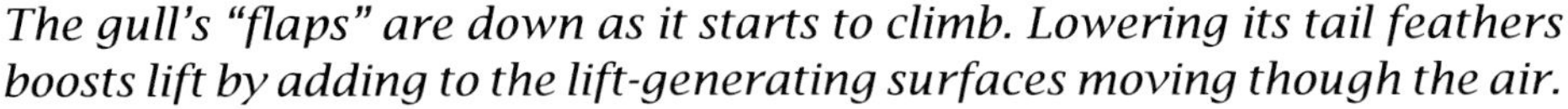

The gull's "flaps" are down as it starts to climb. Lowering its tail feathers boosts lift by adding to the lift-generating surfaces moving though the air.

CHAPTER 6

UP, UP AND AWAY

Takeoff is ultimately a matter of lift. To become airborne, a bird must arrange for the lift generated by its wings to exceed its weight. This can be a formidable task for a bird that is standing still, because lift is created when an airfoil moves *through* the air.

Most bird species fly at cruising velocities ranging from 15 to 45 miles per hour. At these speeds, the airflow past their wings generates ample lift. During level flight, the bird's muscles have the comparatively easy duties of overcoming drag and sustaining the bird's airspeed.

But at takeoff, wing power must deliver all of the energy required to raise the bird's weight into the air. The motion of flapping wings is the chief source of a bird's lift—assisted perhaps by the bird jumping into the air, running along the ground or aiming into the wind.

Takeoff is the most strenuous phase of flying. As a rule of thumb, the larger the species, the less capable it will be of short takeoffs and steep climbs. Large birds, like large aircraft, usually take to the air slowly.

Birds typically make use of three techniques to produce additional lift when they take off:

- They flap their wings faster during takeoff than during level flight (because increasing the speed of an airfoil through the air generates more lift).
- They position their wings high above their head at the start of the downstroke (because a longer duration downstroke produces more lift and thrust).
- They "drop their flaps" to produce extra lift.

The sequence camera caught this herring gull an instant before it became airborne. Its feet are rocking forward, about to leave the ground.

Some birds (pigeons, for example) invert their hands on the upstroke to produce additional thrust.

These variations do not change the fundamentals of flapping flight. You will recognize the familiar stages in the wingbeat—executed much more energetically—in the six-picture takeoff sequence presented inside this double gatefold.

3

The gull's wings are fully extended—and as high above its head as the bird's shoulder joints will permit. The primary feathers are tightly overlapped to create a "solid" surface. Because the "propellers" are located at the ends of the wings, they will move fastest and farthest through the air when the wings swing downward.

2

The gull has drawn its wings way back to maximize the duration of its downstroke. Its hands are pivoting to unfurl its "propellers."

4

The downstroke begins. The gull's "propellers" are angled sharply downward to generate the high initial thrust necessary to launch the bird.

Chapter 7

Some Finer Points of Bird Flight

Birds, like airplanes, can be more fun to watch when they are doing the unusual: taking off, landing, quick turns, mid-air acrobatics. These maneuvers demand that lift and thrust be precisely controlled. A pilot does the job with a stick, a throttle and rudder pedals; a bird changes the shape and orientation of its wings.

A bird's wing is designed to produce lift and thrust simultaneously. The inner wing generates lift throughout the wingbeat, because it remains relatively level during flapping flight—even at the start of the downstroke and the beginning of the upstroke.

The bird's "propeller" produces a force that points slightly uphill during level flapping flight. In other words, the two handfuls of primary feathers make lots of thrust and a little lift during the downstroke. (Hardly any forward thrust is generated during the recovery stroke; the bird's momentum keeps it moving ahead—with little loss of air speed until the next downstroke.)

Many species of small birds take advantage of momentum to conserve energy when they fly. They "bound" though the air, alternately flapping their wings, then coasting with wings folded.

Many larger birds, including crows, owls, gulls and eagles, use a kindred technique called *undulating flight.* It consists of periods of gliding flight followed by spurts of flapping to regain the altitude lost while gliding.

A bird can vary the *pitch* of each "propeller"—the depth of its bite into the air—throughout the downstroke, to control thrust. Most birds steer

This mallard hen is executing a sharp turn to the right by making a one-sided downstroke. The bird's left wing will generate more thrust than its right wing, spinning the duck in the air.

while airborne by boosting the thrust produced by one wing and decreasing (or even reversing) the thrust from the other. They turn with impressive agility—and amazing alacrity.

Bird Wings

The design of all wings—bird or aircraft—involves many compromises and trade-offs. For example, longer wings will generate more lift than shorter wings, but they are also more difficult to flap and fold. And so,

A mallard drake, its wings in their rearmost position, is starting a new downstroke. Ducks have relatively short, stubby wings, capable of powerful, rapid flapping that can achieve near-vertical takeoff and climb.

there are as many different kinds of wings as there are different species of birds. Here are a few familiar rules of thumb about wing design:

- Woodland birds that must maneuver among trees and branches, but which do little long-endurance flying, tend to have short, rounded wings.
- Birds that soar over open water, migrate long distances or spend lots of time in the air usually have long, pointed wings that make less *induced drag* (drag that accompanies the production of lift).

This great horned owl is climbing; it has just completed a powerful downstroke that pulled its wings forward of its head. Many birds cowl their heads during the recovery strokes of high-power wingbeats.

- Birds that spend their days making "lazy circles in the sky" typically have wide wings, equipped with "slotted" wing tips that control tip vortex during slow-speed flight.
- Birds that earn their livings by diving underwater have small wings that do not impede swimming, but which limit their owners to slow climbs after long takeoff runs.
- Fast-moving birds tend to have short, stubby wings that generate plenty of "propeller" power and less *profile drag* (drag that is

The great horned owl's broad tail can "extend" the trailing edges of its wings—like flaps on an airplane—when more lift is needed during climb. Here, the bird is about to flip its hands up to complete a recovery stroke.

produced when an object is pushed through the air), but which have relatively poor gliding ability.

Bird "Propellers"

Leonardo da Vinci was one of the first scientists to observe that the wing tips of a bird in flapping flight—viewed from the side—appear to move in a lopsided figure eight. But the full significance of a bird's hand motion was not understood until late in the 19th Century.

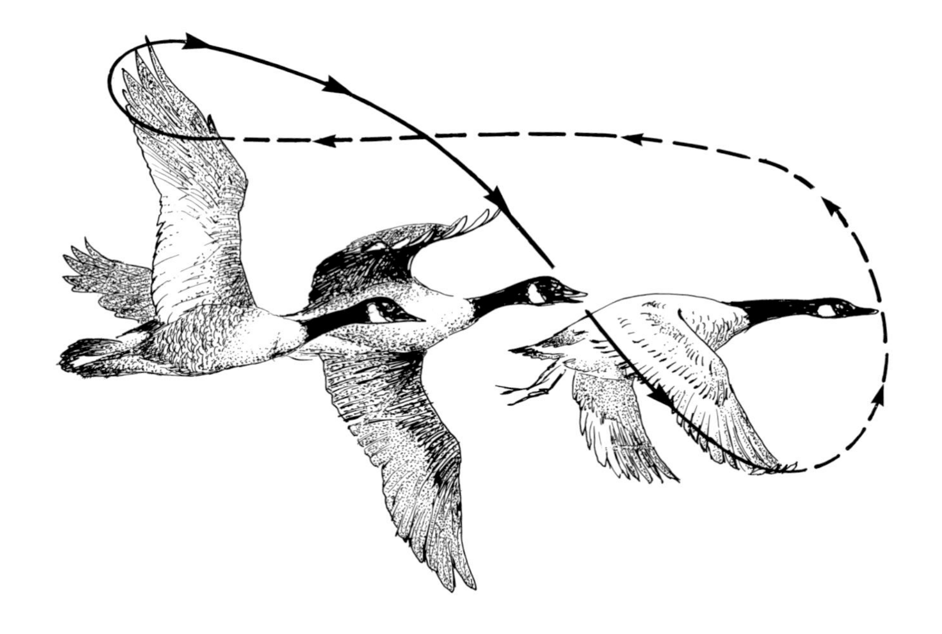

It can be difficult to visualize the moving figure-eight path of a bird's "propeller" as its hand alternately drives downward through the air then folds against the inner arm for the recovery stroke. This illustration greatly simplifies the actual hand motion.

The blades of an airplane propeller generate thrust continuously as the prop rotates in a full circle. The process is easy to visualize and comprehend. (Propellers are sometimes called *airscrews*, a nicely descriptive label.)

A bird's "propeller" follows a much more puzzling path through the air. Seen from the front, the downstroke begins with a half-circle swing from above the bird's head to below the bird's body. But then comes the recovery stroke—with its complicated chain of smoothly syncopated movements that happen in the blink of an eye:

- The elbow joint starts to flex and the inner wing begins to rise—before the primaries have finished biting into the air.

A gannet, bringing some building materials home, is flying at low speed, with its "flaps" down. The bird is executing a recovery stroke. It has begun to rotate its hands downward.

- The wrist joint flexes; the hand rotates downward against the forearm as the wing swings high above the bird's head.
- The "propeller" snaps open with a flick of the bird's wrist.

In fact, there is a simple purpose behind this apparent complexity: By rotating its hand, the bird eliminates much of the air resistance that might otherwise slow its recovery stroke. (Think of paddling a canoe; if you can't lift the paddle out of the water at the end of each stroke, your next-best choice is to turn it sideways.)

The gannet's hands are rotated downward to minimize air resistance as it pulls it wings back to complete the recovery stroke.

With a flick of its wrists too quick to catch, the gannet rotates its hands "open" and begins the next downstroke.

CHAPTER 8

DRUMMING INSTEAD OF FLYING

A ruffed grouse cock *flies* to lure a mate, but he never leaves the ground.

This apparent paradox has a simple explanation: the famed drumming sound produced by the male ruffed grouse is the noise of a wingbeat. For that reason, I have included grouse drumming in this book.

A ruffed grouse cock is a highly territorial, non-migratory bird. Each spring he stakes out the best woodland territory he can find, then begins to drum. Drumming is part of his behavior to establish and maintain his exclusive territory and defend his turf against rival males.

Drumming is also the way a ruffed grouse cock entices a mate to his wooded dominion.

After the female is on the nest, the cock continues to drum to assert his claim to his territory. He will strut about with ruff raised, shake his head and hiss if another cock appears.

Grouse typically choose a favorite "drumming log" in their territory and return to it whenever they drum. One theory holds that standing on a log elevates the grouse so that he can see farther—presumably to get a good view of rival males or potential mates. (Clearly, the grouse on these pages found a "drumming rock" perfectly satisfactory.)

Drumming starts with measured thumping sounds that resemble the muffled thuds of an old tractor engine. Then the thump rate accelerates quickly in tempo, until the beats commingle into a whirring noise that lasts for eight to ten seconds.

The drumming sound is audible for long distances in the woods, but is hard for a human ear to pinpoint. Grouse ears have no such difficulty:

The "downstroke" of the grouse's drumming wingbeat makes noise but does not create lift. The bird raises its ruff when it drums to attract a mate and to defend his territory against rival males.

A cock grouse will readily zero-in on a rival male who has the temerity to drum in its territory.

The grouse is a chicken-like bird that usually weighs less than one-and-a-half pounds and has a wingspan of 22 or 23 inches. Nevertheless, a drumming grouse cock can produce an impressive "prop blast" with its wings. Dry leaves or pine needles on the forest floor around the bird are swept away by the airflow.

The drumming wingbeat is strong and incredibly rapid—it appears completely blurred to the human eye, rather like a fast-turning fan. So it is not surprising that the source of the drumming noise was long debated.

Does the bird pound its wings on the log?

Does a grouse clap its wings together like a pigeon?

A drumming grouse swings its wings behind its body to complete the "upstroke" of its wingbeat. The upstroke and downstroke must displace equal volumes of air, because the bird stays put as it drums.

Is the sound generated by special vocal cords that are activated by air racing from the bird's airsacs as it beats its wings?

The uncertainly lasted long after the invention of photography, because wild grouse are among the most uncooperative models in the avian kingdom. The first sequential photographs that show the drumming wingbeat in sufficient detail to explain its mechanics were taken in the late 1920s.

The grouse drumming on these pages are wild birds. I took their pictures with a radio-controlled camera and electronic flash, while I stood some distance away.

A drumming grouse holds its fan of primary feathers open and aligned with the direction of its wingbeat. Consequently, drumming does

The grouse's broad tail probably helps stabilize his body at the start of drumming. Note that the secondary feathers on the bird's inner wing are also spread apart, presumably to reduce air resistance.

not produce enough forward thrust to pull the bird off its perch. (The secondary feathers on the inner wings also seem to be spread apart slightly to reduce air resistance.)

The grouse swings its wings back and forth through the air; they do not produce lift because hardly any air moves *across* their airfoil surfaces when the grouse drums. The "upstroke" and "downstroke" apparently shift equal volumes of air—which leaves the grouse firmly planted on his feet.

As the wings bat the air, they create the distinctive low-frequency drumming noise. Some scientists explain the sound as a sequence "miniature sonic booms" produced by the air that rushes in to fill the space vacated by the fast-moving wings.

INDEX